Fresh Food Science

A Coloring Book

By Dr. Megan Olivia Hall, PhD
Illustrations by Liz Carlson Tomczak

Cedar Bend
Books LLC

Written by Megan Olivia Hall
Cover art by Liz Carlson Tomczak
Cover design by Paloma Leone-Getten
Illustrations by Liz Carlson Tomczak
Layout and design by Paloma Leone-Getten

Cedar Bend Books LLC
St. Paul, MN
www.cedarbendbooks.com

Library of Congress Cataloging-in-Publication data are available.

ISBN (paperback) 979-8-9915656-1-5
ISBN (ebook) 979-8-9915656-3-9

First Edition

Table of Contents

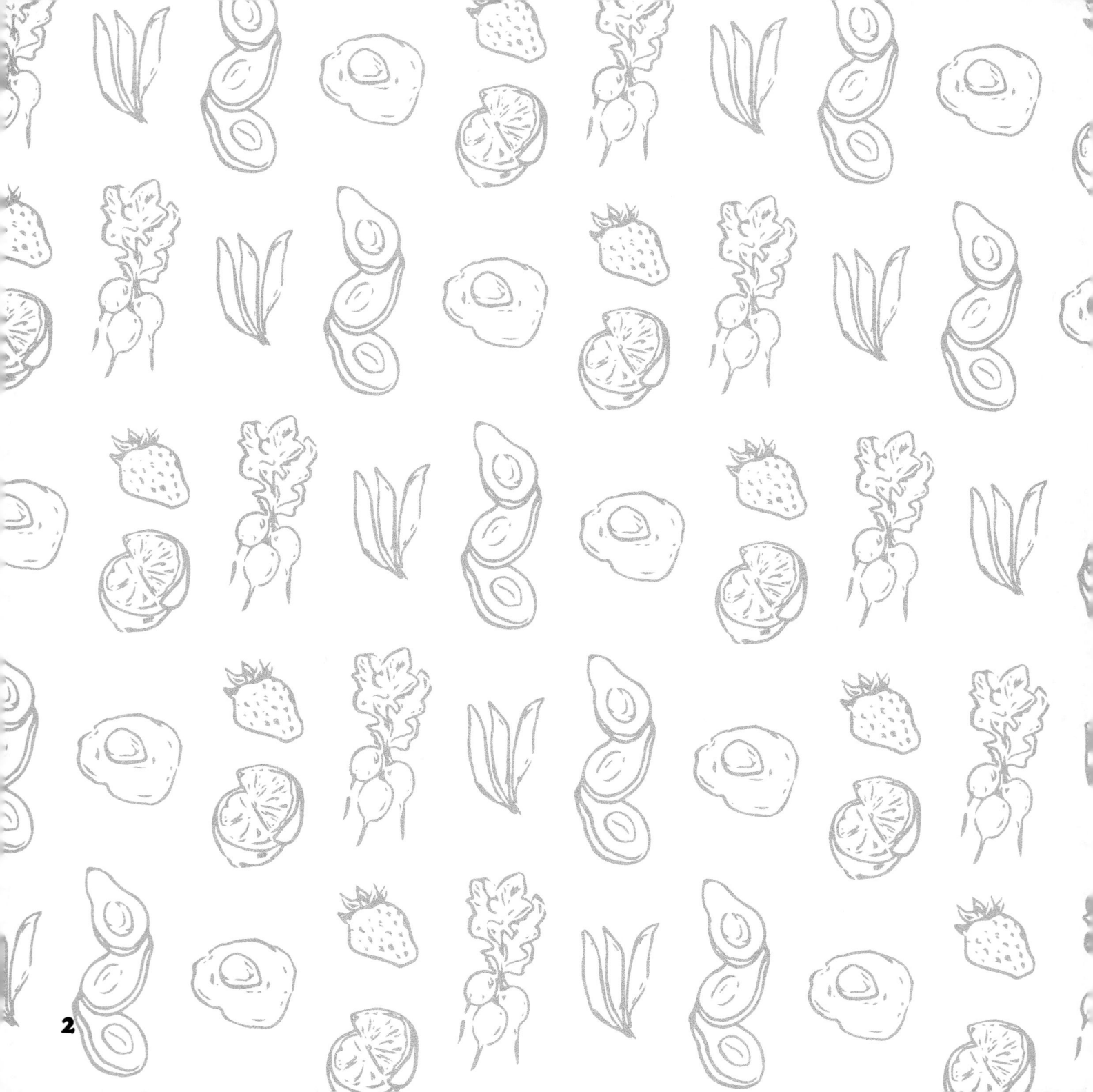

Introduction

Welcome to *Fresh Food Science: A Coloring Book!* I hope you enjoy relaxing with 39 of Liz Carlson Tomczak's beautiful line drawings from *Fresh Food Science: 101 Healthy, Easy, Delicious Recipes for Tween, Teen & Young Adult Chefs*. The single-sided pages in this book work wonderfully with watercolor markers, other water-based markers, alcohol markers, brush pens, and colored pencils.

Welcome to the Kitchen!

While cooking from scratch may take time and attention, eating freshly made food means having more energy for all the things you love to do in life. You are worth the effort for homemade, delicious, healthy meals. Learning how to cook well is one of life's greatest joys.

Washing Produce

When cooking with fruits or vegetables, be sure to wash your produce before you begin cooking. The safest way to wash fruits and vegetables is to rinse them with water.

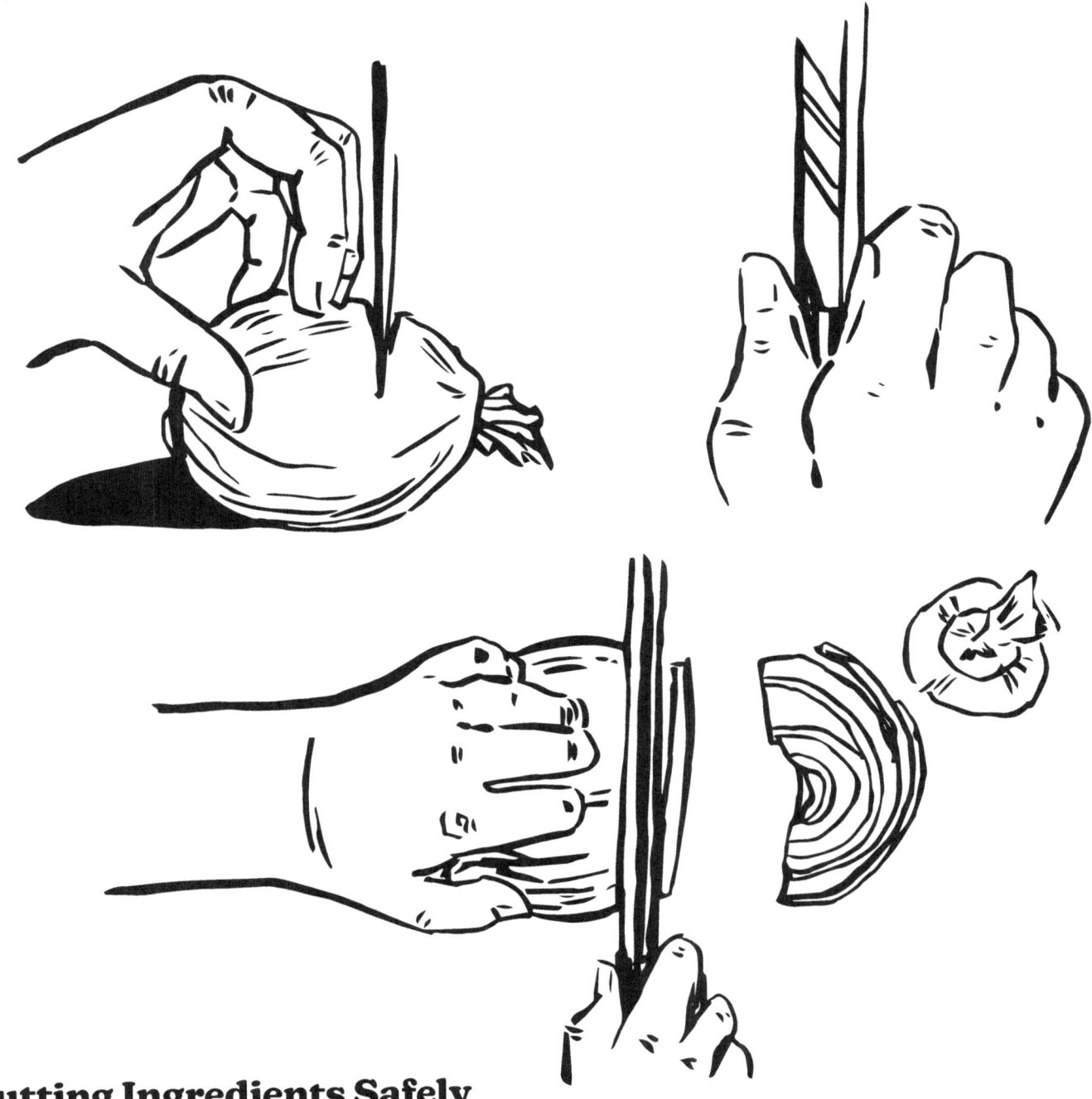

Cutting Ingredients Safely

When cutting, move the knife, not the food. Tuck your fingertips inwards so that the blade is near your knuckles, not your fingertips. Immediately after using a knife, clean it, dry it, and put it away.

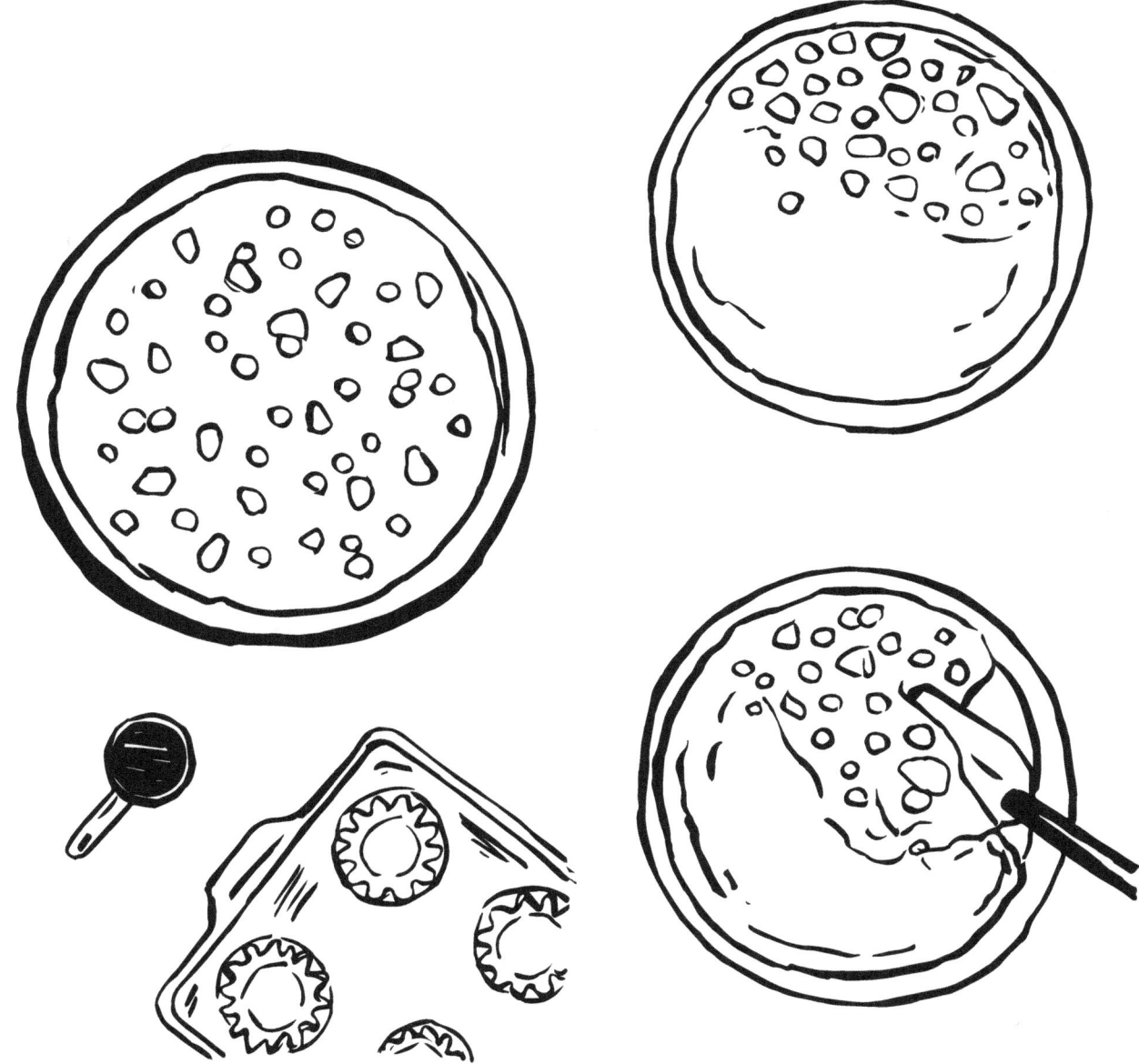

Combining Ingredients

When combining, or mixing, ingredients, there are a variety of techniques to use. Folding is the gentlest mixing method and should be used to keep air bubbles intact in a light and fluffy mixture. To fold ingredients together, use a rubber spatula and softly stir lighter ingredients into heavier ones.

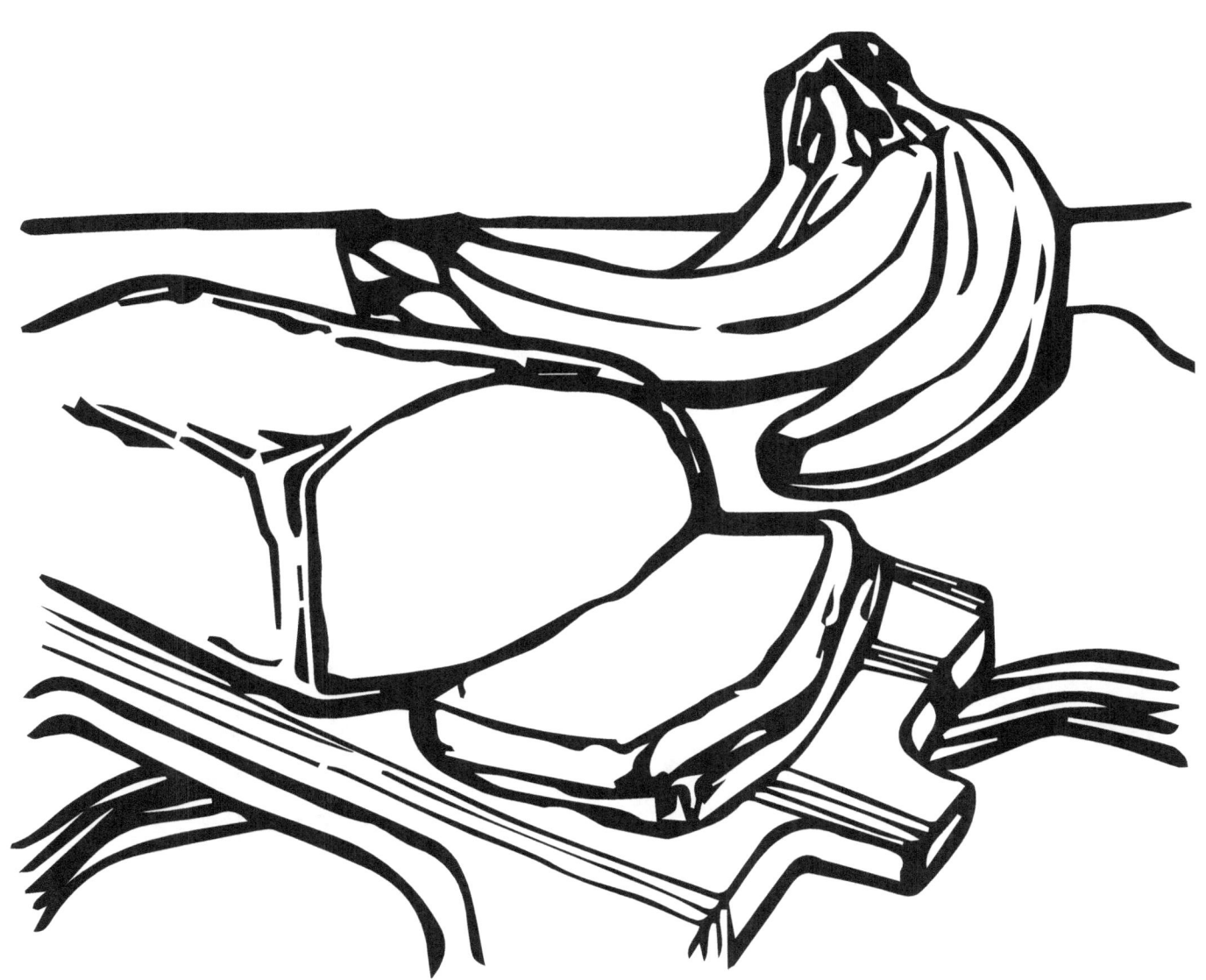

The Whole Banana (Bread)

The deep heartiness of buttery, whole-wheat banana bread is sweet with honey and tangy with yogurt. Enjoy this treat with a hard-boiled egg and an apple for a balanced breakfast to go.

Kitchen Sink Granola

Homemade granola has fewer sweeteners, healthier oils, and can be customized to include whatever add-ins you love most. When making your own granola, try many different add-ins to discover what you like best.

Order Up Oats

Hot porridge topped with fruit makes for a cozy and comforting breakfast that cheers up even the dreariest winter Monday. Cold overnight oats are a great hack for busy mornings.

18

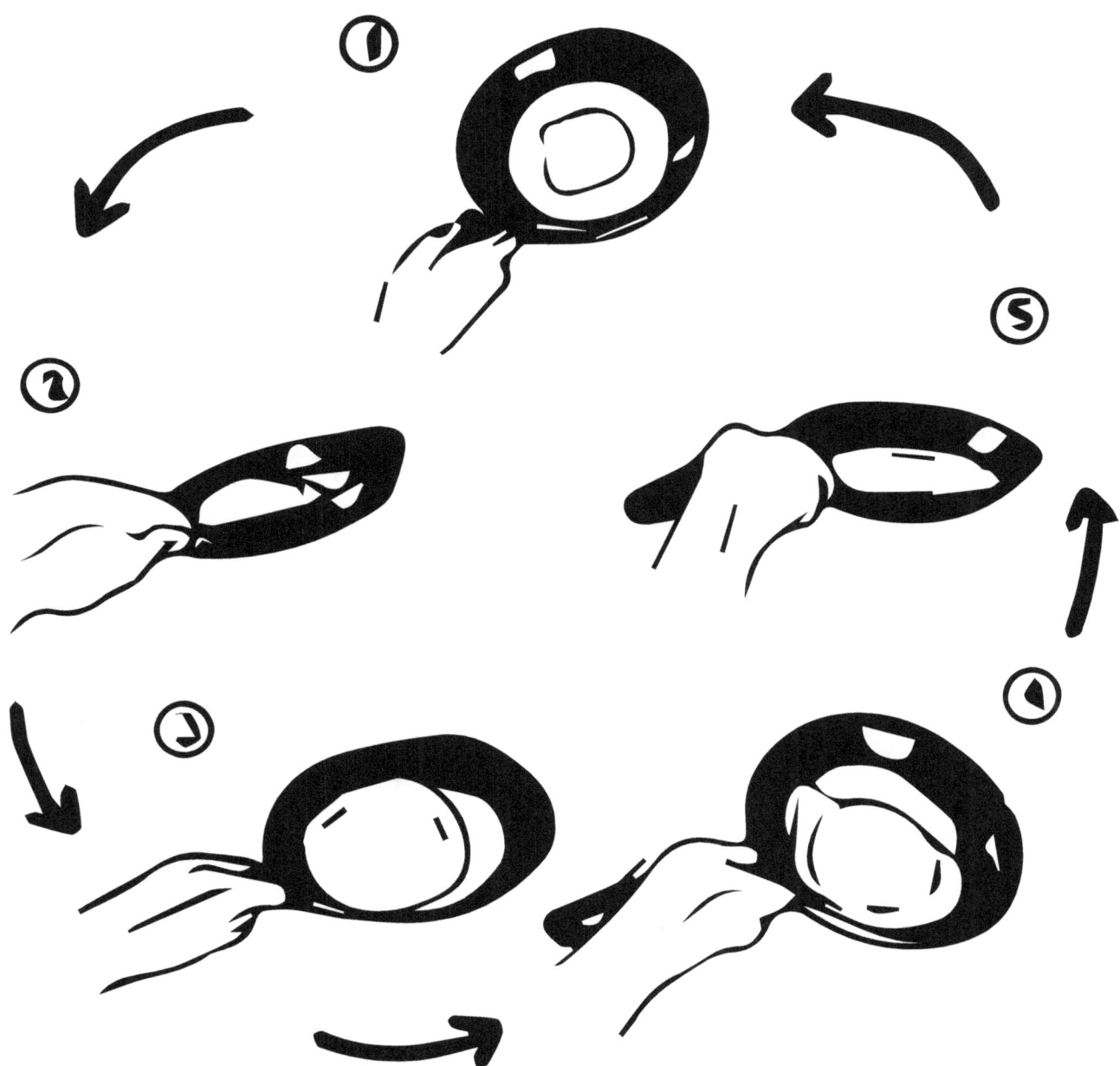

Can-Do Crêpes

To form a crêpe out of batter in a hot skillet, use an oven mitt to quickly pick up the skillet by its handle. Tilt the skillet in a gentle circle until the batter coats the whole skillet in a thin, even layer.

Fluffy Egg Bake

Eggs work as leavening, or ingredients that make baked goods rise. When whisked, egg whites form a durable foam that holds air bubbles even during baking. Air bubbles expand in the hot oven, creating a risen bake with a fluffy texture.

Feel-Good Frittata

Vegetables for breakfast?! Baked frittatas are a super easy way to load up on nutritious veggies (and protein!) in the early hours of morning – or at night, if you like having breakfast for dinner.

Sandwich Solutions

Whole wheat focaccia bread is a flavor explosion! The whole grains, olive oil, flaky salt, and rosemary are a delicious solution to elevate any sandwich filling.

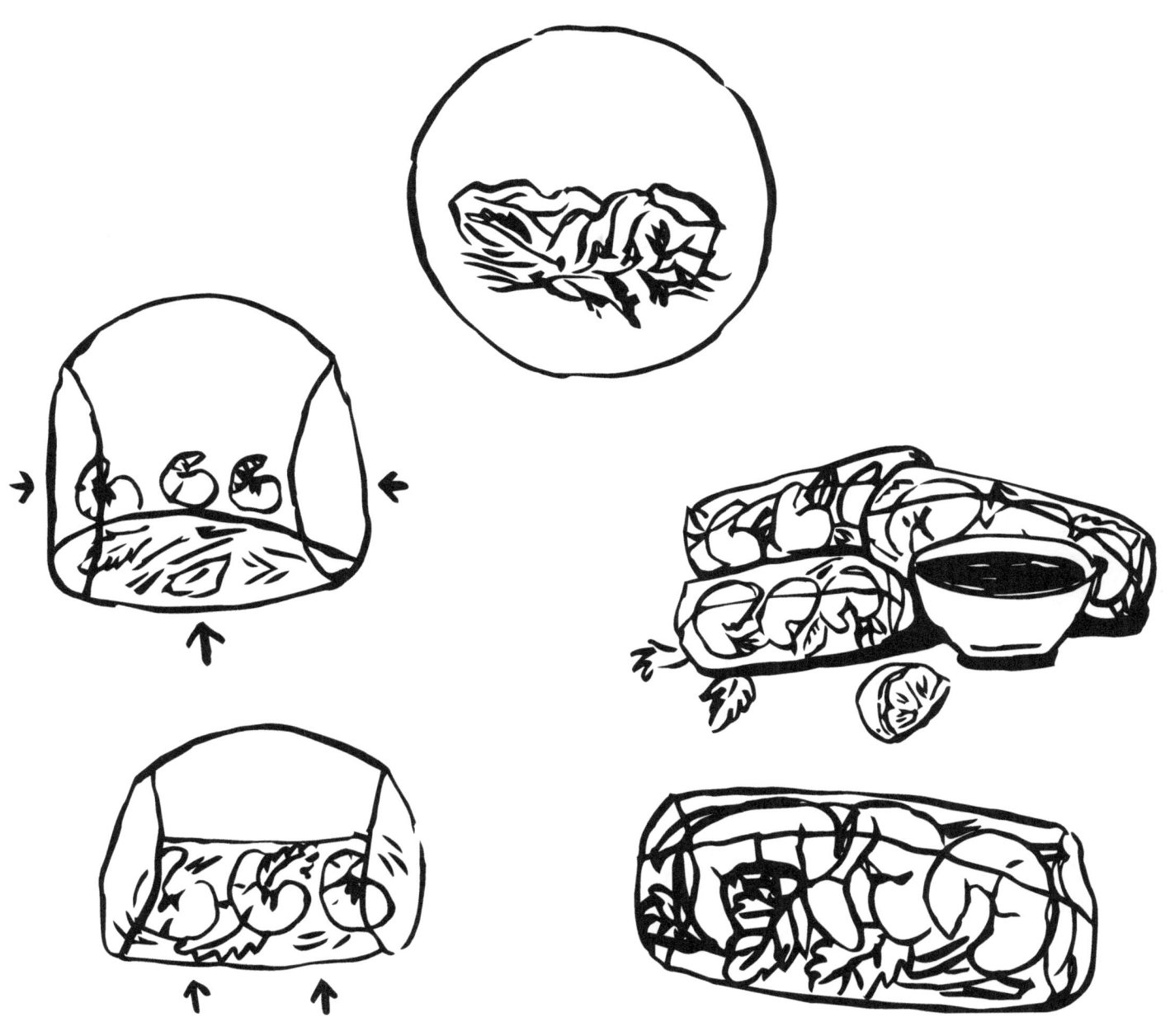

Mango Spring Rolls

Spring rolls are fresh, bright, hand-held salads. Mango adds a smooth sweetness that plays well off the crisp flavors of mint, basil, and cilantro.

Next-Level Ramen

Instant ramen can become a healthy gourmet treat when leveled up with fresh vegetables and savory seasonings.

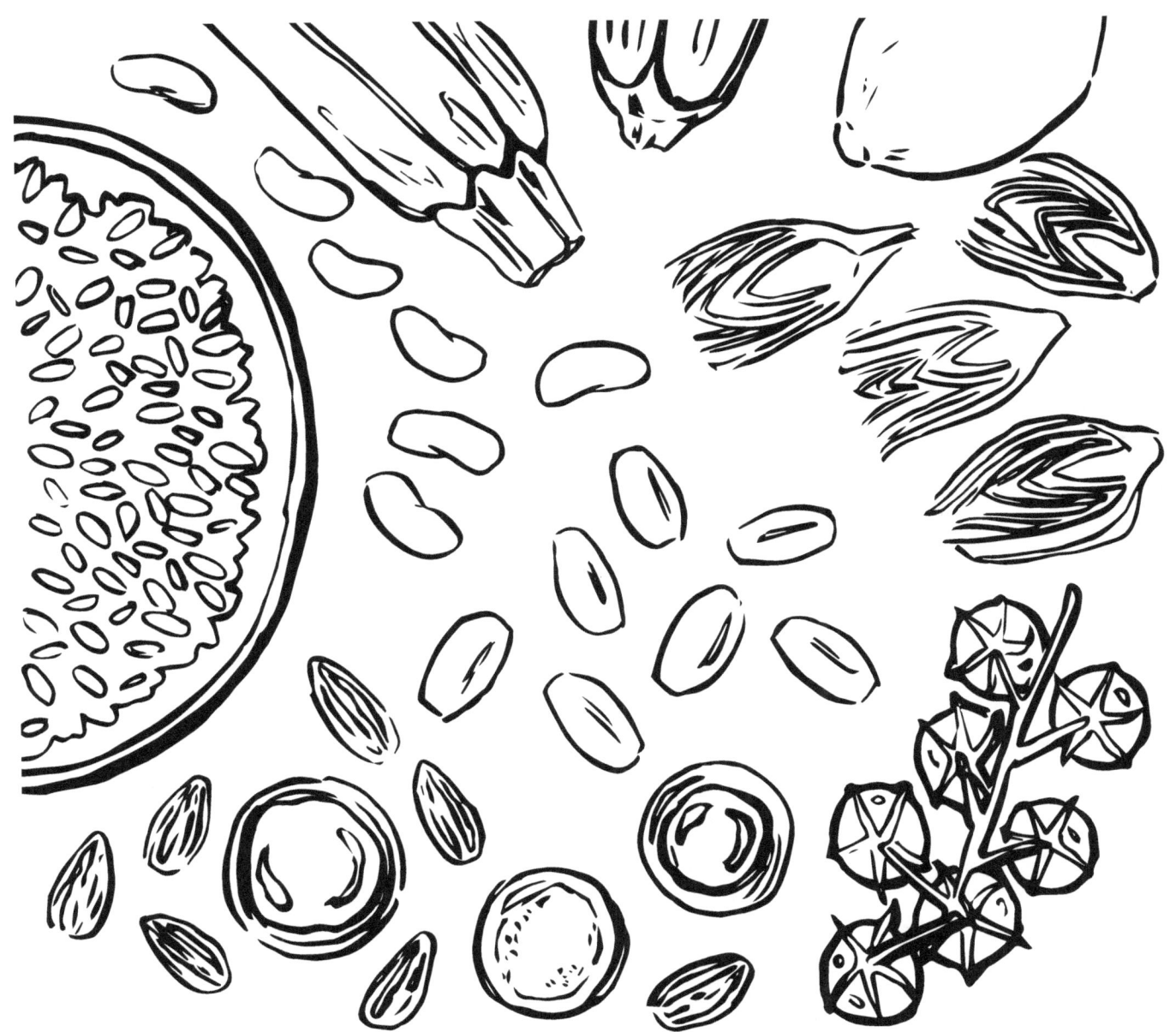

Lemony Pasta Salad with Disappearing Zucchini

Grated zucchini has a similar texture to cooked orzo, so it easily disappears into this salad. But zucchini aren't the only sneaky fresh foods here. Artichokes usually pass as vegetables, but they are actually flower buds. These vitamin-packed imposters, if left to grow, turn into 3-foot-tall purple flowers!

Snack Packs

A few minutes of planning on shopping day can make packing a healthy lunch – even when you're in a rush – super simple. Grab a bento box style lunch container and fill the compartments with a whole grain, vegetable, protein, fruit, and dessert.

Leftovers for Lunch

For the quickest possible lunch, pack up leftovers from a recently cooked meal. Make sandwich with Bison Meatloaf or Hearty Egg Bake, layer The Whole Banana (Bread) with peanut butter, or keep Three Sisters Chili warm in a thermos.

Fresh Out of the Fridge

Snacks don't have to be processed or packaged. Try setting fresh foods like apples, yogurt, or sugar snap peas on your fridge's center shelf so they're the first thing you see when you start rummaging for a bite to eat between meals.

Go-Go Gorp

Gorp is satisfyingly sweet and salty at the same time. The sweetness comes from natural fruit sugars. The richness of dark chocolate and the healthy fats in salted nuts make gorp a filling snack.

H₂O+

Because water carries oxygen to all cells, and because cells are the building blocks of the body, drinking plenty of water makes your whole body stronger and more energetic. From your blood to your brain, from your muscles to your metabolism, everything in your body works better when you drink 6 to 8 cups of water every day.

Quintessential Quiche

The key to a flaky quiche crust lies in the layering of ingredients. Cheese is high in fat. Because fat and water don't mix, a layer of cheese at the bottom of the quiche creates a waterproof barrier that prevents the vegetables and custard from soaking into the crust.

Potato Leek Soup

Scientifically, leeks are classified as alliums, along with onions, garlic, scallions, chives, and wild ramps. Allium vegetables contain antioxidants including allicin, a powerhouse chemical that fights infections, lowers cholesterol, and prevents cancer.

Lasagna Soup

Just like traditional lasagna, most of the delicious but relatively unhealthy sodium and saturated fat in this recipe come from cheese. In traditional lasagna, much of the cheese disappears into hidden layers. In lasagna soup, ricotta and parmesan melt over the top of the soup, giving maximum sensory enjoyment for every ounce of cheese.

Sesame Ginger Stir Fry

Layered with bright, rich, salty, and umami flavors, stir fry is a fun way to load up on vegetables. Bell pepper, broccoli, sugar snap peas, cauliflower, bean sprouts, cabbage, celery, tomatoes and zucchini are all delicious in stir fry.

Pizza Party

Ever wonder why melted cheese tastes sooooo good? Heat relaxes the protein matrix holding cheese together, releasing rich, creamy fat and tasty, umami-flavored amino acids. Heat also releases as many as 50 new scents from cheese, which activate the olfactory, or smelling, part of the taste experience.

Fresh Fried Rice

Fresh fried rice is a flexible recipe that can absorb any leftover vegetables in your crisper drawer. For the best flavor, be sure to use at least one allium, like an onion, shallot, garlic, or green onion.

Hotdish Revival

In Minnesota, we call casserole *hotdish*. On a snowy winter night, hotdish is the ultimate comfort food. Traditional hotdish is made with canned condensed soup. To keep it fresh, substitute a simple roux made from scratch.

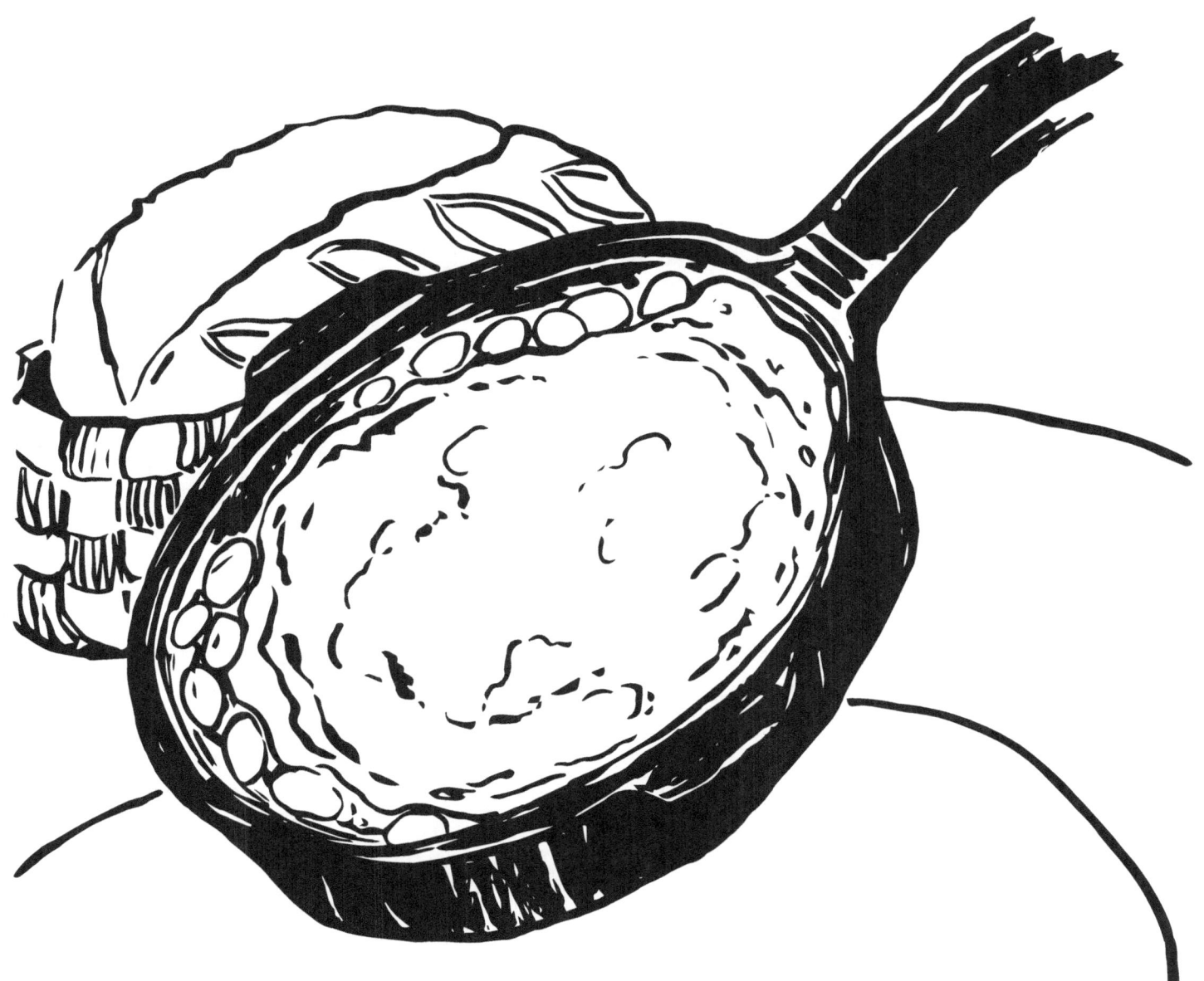

Easy Cheesy Bean Bake

Cannellini beans are white beans, like navy beans, lima beans, and Great Northern beans, which are full of fiber, protein, and antioxidants that protect against chronic illnesses like heart disease and cancer. When shopping for beans, be sure to check the nutrition label for low- or no-sodium options.

Chia Pudding 5 Ways

Chia seeds are very rich in fiber, healthy omega-3 fats, vitamins, minerals, antioxidants, and protein. Chia seeds also contain mucilage, a chemical common in many plants that traps water. When chia seeds are sprouting, mucilage ensures that the baby plants have enough water to grow. In chia pudding, mucilage absorbs the water in cow, oat, soy, almond, or coconut milk, thickening the pudding with its now-gooey texture.

Happy-oca Pudding

Tapioca is made from the root of a plant called cassava (also known as mandioca and yuca). Originally grown in Central and South America 5,000 years ago, tapioca is now an important part of many cuisines around the world. A gluten-free starch that naturally thickens puddings, and stews, tapioca is also the main ingredient in boba pearls.

Carrot Cake with Whipped Cream Cheese Frosting

The secret to smooth, fluffy cake and frosting lies with great mixing technique. Mixing fats like oil, butter, or cream cheese with sugar dissolves the sugar (preventing grittiness) and builds in tiny air pockets. Egg protein is especially good at holding air pockets, so it's important to keep mixing the batter after adding eggs. In the hot oven, the air pockets will expand, lifting and fluffing the cake.

Honey Apple Cinnamon Cake

Honey, made by honeybees from the nectar of flowers, is an incredible substance, full of antioxidant, anti-inflammatory, and antibacterial compounds. While visiting flowers, bees pollinate about 75% of the fruits, nuts, and vegetables grown on U.S. farms.

Yeehaw Cookies

Ever notice that chocolate chips keep their shape even after baking in a hot oven? Chocolate chips contain food additives called stabilizers to help them keep their shape and not melt into puddles of goo. *Fresh Food Science* recipes always call for chocolate baking bars to avoid these additives.

67

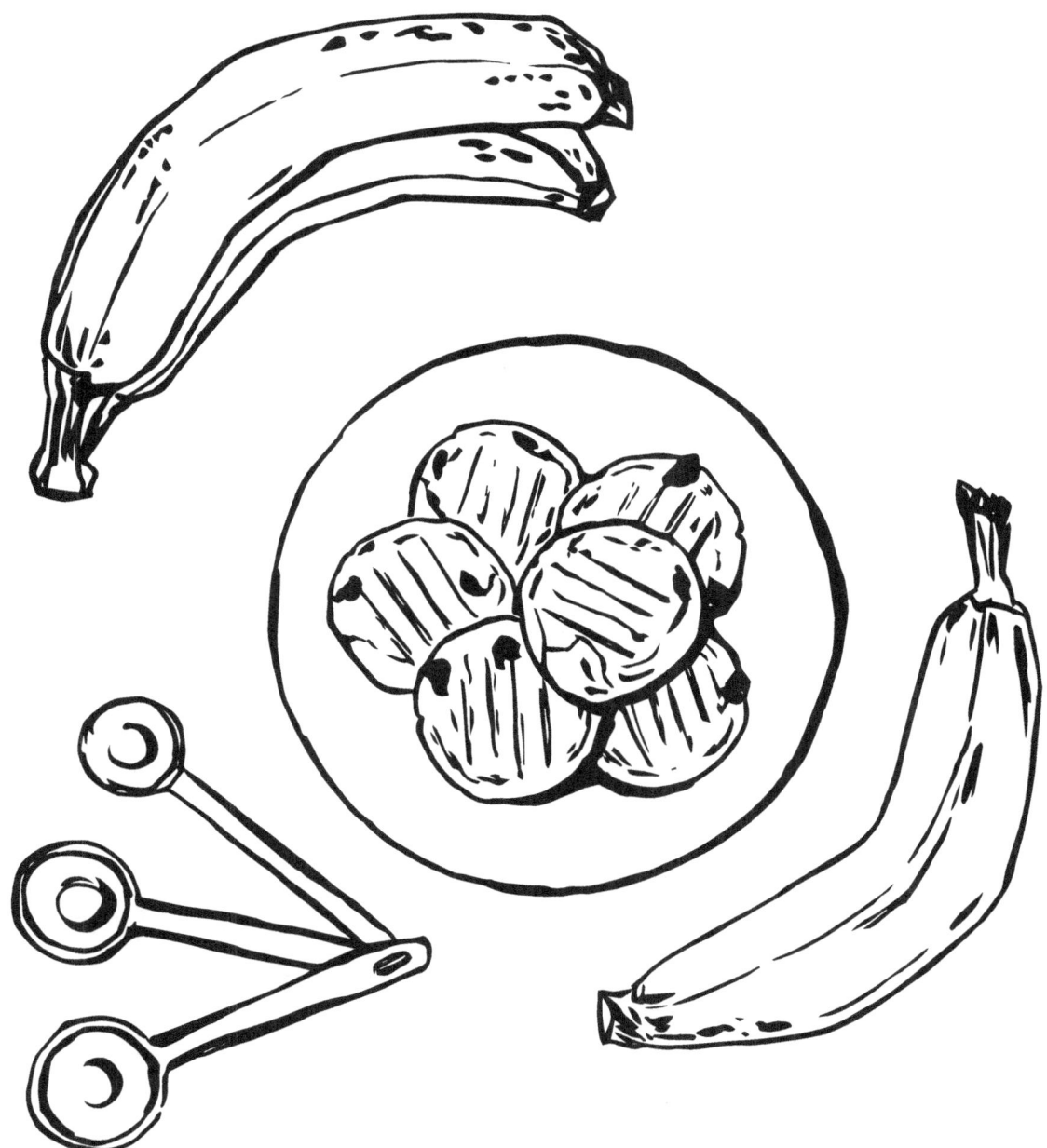

Banana Chocolate Chunk Cookies

Almond flour, although lighter in mass than wheat flour, is higher in fiber, protein, and fat. Butter is not needed in these cookies, which get their rich, toasty flavor and tender texture from almond flour.

Figgy Date Energy Cookies

Cashews, figs, and dates are rich in antioxidants, fiber, and copper, a mineral needed by the brain, the immune system, and for the body to generate energy from food. Cashews have just as much protein as meat but are healthier for the blood and heart.

Energy Boost Biscotti

Biscotti, in Italian, means twice baked, and while you can pop these cookies into the oven a second time after slicing for a crisper texture, they are quite good after just one baking.

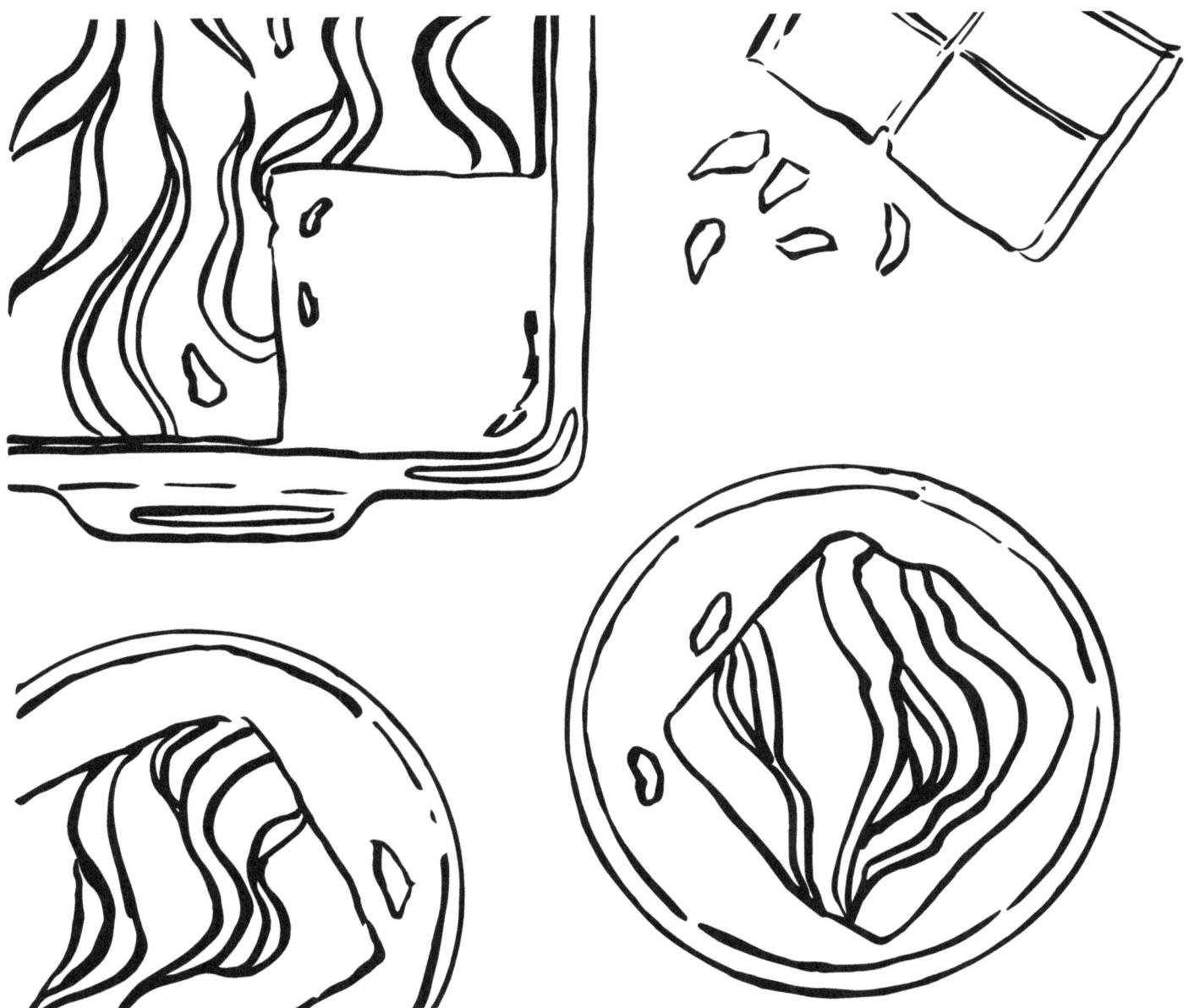

Peanut Butter Black Bean Brownies

Black beans bring protein, complex carbohydrates, fiber, and a creamy texture to brownies while lowering their overall fat content. Once pureed, black beans have a dough-like texture that helps them disappear into the brownie batter.

Blueberry Cardamom Cobbler

Coconut milk and cream, which are rich in fat, can be used as vegan substitutes for butter, milk, buttermilk, yogurt, or eggs in most baking recipes. Because plants don't make cholesterol, coconut milk does not contain this artery-clogging nutrient.

Fresh Ingredients

The key to preparing fresh, healthy meals is to start with fresh, whole ingredients like fruits, vegetables, and eggs. Scientifically, avocados and lemons are considered berries because each fruit grows from a single flower. According to plant scientists, strawberries are multiple fruits, made of many small fruits together. Strawberries are not berries at all!

Kitchen Equipment

A well-equipped kitchen is stocked with essential appliances and tools. What's your favorite kitchen accoutrement?

About the Author

Megan Olivia Hall is a STEAM educator and instructional guide at Open World Learning Community (OWL) in St. Paul Public Schools. In her 27 years of teaching, Megan has worked with students of many ages and levels, from preschoolers to graduate students. A National Board Certified Teacher, she is the 2014 Minnesota Teacher of the Year and a 2015 NEA Foundation Global Fellow. Megan regularly presents at national teacher conferences. She consults through SPARK Teaching Group, leading professional development for teachers in the areas of team building, supporting student mental health, and teaching with joy.

Megan holds a Ph.D. in Learning, Instruction, and Innovation from Walden University. Megan's writing has been featured in Education Week and The Science Teacher. Her books include *Awesome Kitchen Science Experiments for Kids*, an Amazon #1 Best Seller in Scientific Experiments & Projects, as well as *Awesome Outdoor Science Experiments for Kids*, *Big Chemistry Experiments for Little Kids*, and *Adventure Girls! STEM Crafts*.

Megan lives in St. Paul, Minnesota, with her husband, two kids, and three cats. When she's not concocting new recipes (or old favorites) in the kitchen, she enjoys canoeing, gardening, reading, and yoga.

About the Illustrator

Liz Carlson Tomczak is an artist, illustrator, and teacher in St. Paul, Minnesota. She works in a variety of mixed media ranging from colored pencil to gouache and beyond. After working two decades in the art materials industry, she is a National Certified Art Materials Expert. She loves sharing her passion for making art with her students of all ages. Her favorite subjects to recreate are food, pets and then everything else. She shares her artist's loft home with her husband, a retired chef and their terrier mix, Noodles. See more of her art on Instagram at @lizcarlsonart. This is her first project with Cedar Bend Books.

About *Fresh Food Science*

Fresh Food Science: 101 Healthy, Easy, Delicious Recipes for Tween, Teen & Young Adult Chefs is a fun, welcoming invitation for kids, tweens, teens and the adults in their lives to invest a little time to get a lot of great-tasting, healthy meals. Encouraging, clear, and simple instructions help young chefs learn to navigate the kitchen. Each recipe includes an engaging watercolor or line drawing illustration, nutrition facts, and a food science connection backed by credible resources cited in the recommended reading section. Recipes emphasize cooking with fresh foods, including fruits, vegetables, whole grains, and minimally processed proteins while reducing the use of refined sugars and flours.